The Federation of Blue Planets

A heartwarming story of a 25th Century professor's reunion through time with his 21st Century family ancestor.

Copyright 2016 by Author

All rights reserved, including
the right to reproduce this book,
or portions thereof, in any form.

ISBN-13 #978-1541263185

Author's Prologue

THE STORY OPENS in celebration of a historic occasion, Earth's inauguration into The Federation of Blue Planets. Earth has become a virtual paradise as the result of global peace and the tremendous advances in science and technology.

Included in the millions of planets inhabited by interstellar traveling civilizations throughout our galaxy, thousands of them are Earth-like planets.

With the aid of a more advanced member of The Blue Planets, Derrick Patton is able to travel through time and pay a personal visit with his ancestor, Professor Robert Berg, in the 21st Century.

MY NAME IS Derrick Patton. I am a Professor and Astrophysicist at Earth's leading Interplanetary University of Science and Space Exploration in Albuquerque, New Mexico.

It is my hope that future generations of my family, and others, will refer to my journal, and cherish it. Just as I have cherished a family diary passed down to me from the year 2137.

How profound it would be if they could see how it all turned out!

This LOG is my personal journal.

LOG 438421

April 20, in the Year 2472

I HAVE BEEN invited to be a guest speaker to celebrate a historic occasion. Everyone on Earth and many others beyond our planet will be watching our formal inauguration into The Federation of Blue Planets. We are already members of the Galaxy Alliance of Planets which is inclusive of all interstellar space traveling civilizations.

The focal point of my lecture will bring together the past four and one half centuries into where we are today on this momentous occasion. As a historian, I am well aware of the many crises that our planet was confronted with in the 21st Century. Many times, in the first half of this century, it seemed that we would not survive as a civilization.

The theme of my presentation is one of gratitude to our fellow citizens of Earth in this period of the 21st Century. I am speaking as if this were a message of Good News to be delivered through time. Many in this era played a vital role in bringing about peace between nations, peace that literally saved our planet. One of which was an ancestor of mine, the renowned Astrophysicist, Professor Robert Berg.

Professor Berg gave hundreds of lectures on the subject of Astronomy, Cosmology, and the future of interstellar exploration that included the possibility of time travel. However, his most significant contribution was his unique way in the application of this knowledge for world peace. His profound wisdom, insight, and leadership played a vital role in bringing about the final ending of all international conflicts. His lectures and writings are available within Earth's archives to be read or visually viewed.

THE INAUGURATION LECTURE

by Professor Derrick Patton

A message through time from the 25th Century to the 21st Century

"MY FELLOW COLLEAGUES, beloved citizens of Earth, and to our entire galaxy family. Yes, I call you family because that is the sacred bond which we have established.

"The theme of this lecture is a message of supreme gratitude to those people of the 21st Century who dedicated their lives to world peace. If it were not for them, we would not be here today to witness what an amazing universe this actually is. Please let me take you on an imaginary trip through time to let the courageous people of that era know that we made it in a way beyond anything they could ever envision.

"Let me begin. My dear friends of the 21st Century; this is to let you know that what you accomplished between the years of 2028 and 2057 brought us through and out of that place where a global civilization either makes it or destroys itself. I am telling you this in the year 2472. YOU DID IT! We made it beyond anything you could possibly imagine.

"Today, our Earth is celebrating becoming a member of our Galaxy's 'Federation of Blue Planets'. For more than three hundred years, war and man's inhumanity to his fellow citizens of the Earth has long passed. Medical science has virtually solved all health issues. An unlimited supply of food and energy are now available to everyone. Reports of crime are something that can only be found in our chronicles of history.

"The Blue Planet Coalition will create an open door into the wonders and beauty of hundreds of other Earthlike worlds throughout our galaxy. We have turned our entire Earth's land and oceans into what could be termed a Garden of Eden. Most of these other Earthlike worlds are even more beautiful than ours. We will now be able to visit and experience in person the magnificence of these planets. Most all of them are completely compatible to support all forms of Earth life.

"Many are far more advanced in the field of technology and their knowledge of the universe than we are. The Blue Planets that have especially captivated our attention are the ones that have apparently transcended the science of technology, which has taken them into a realm of existence beyond where technology cannot go.

"Incredible advances in the science of space exploration and travel have been achieved. What your era of science fiction called warp speed has made it possible to transcend the speed of light several thousand times over. Our scientific knowledge has advanced to the point of eliminating the problems that were initially inherent, in what you referred to, as hyperspace travel. We have discovered multitudes of other advanced civilizations throughout our galaxy. We still occasionally refer to our galaxy as the Milky Way, as you did.

"As of this date, we have only been in contact with, and have direct knowledge of, life as it exists in our own galaxy. For the past 200 years, even with our warp speed capability, we have only been able to visit planets and solar systems within a limited area of our local arm of the galaxy sometimes referred to as the Orion Arm. As you know, this is a minor arm in our galaxy located between the two major ones of Sagittarius and Perseus.

"With the many hundreds of solar systems we have explored, we have barely scratched the surface even in our 'neck of the woods' so to speak. With our Orion Arm location being some 3,500 light years across and approximately 10,000 light years in length, you can see the vastness of our own neighborhood in the galaxy. With the overall galaxy being 100,000 light years wide, we have just begun this amazing adventure.

"Within the past 200 years, we have discovered and made contact with hundreds of other planets that are very similar to our Earth. All of these planets present the same beautiful blue color when viewed from space. The atmosphere is basically the same as ours. The climate and environment on many of them are inherently richer and purer than on Earth. We have been able to visit in person only the ones located in our local region of the galaxy.

"However, with the remarkable means of communication that our planet Earth is now capable of, along with the more advanced science in this field of other civilizations, we can communicate virtually in person anywhere within our galaxy that we cannot physically travel. Even though many other civilizations have the capability to visit any part of our galaxy within a very short span of time, there is another approach that in many cases can serve the same essential purpose in what you might say is 'a more convenient way'.

"This is done in a computerized designated area. I am using the term computerized as a point of reference. The technology is far beyond anything related to computers as you know them in the 21st Century. This selected area can be made to appear any place on earth. It is not restricted to or inside of any enclosed structure. Wherever and whomever we are to communicate with are made visible in a three-dimensional, tangible form.

"Any chosen setting from Earth or another world can be provided, including an unlimited number of naturist settings in their highest presentation of beauty. All of the five physical senses are experienced to their fullest capacity within this exact virtual duplication of reality. This can be accomplished almost instantaneously through the means of utilizing one of the dimensions of hyperspace.

"The question may well be asked: why could we not pick up any communication throughout the 21st Century from these thousands of other worlds that were capable of contacting us and were aware of our existence? Well, as many of you in your time had surmised, you were correct. There is a very significant reason for not interfering with any planet's evolution regarding their own overall race consciousness. If this nonintervention is violated, it can significantly affect their ongoing progression and the global maturity required to fulfill the destiny they are to achieve for themselves and for their unique place in our galaxy.

"The explanation of this can best be understood within a metaphysical or spiritual realm. When intelligent life-consciousness evolves on any given planet, there seems to be a unique reason and purpose for its appearing. In other words, there appears to be an intelligent law in operation behind all of this. Seeing the universe as we see it now with all of its thousands of known intelligent civilizations, this has become evident. I will leave the fascinating details of this perspective for another occasion. At the appointed time, when a permanent global peace has been attained on any given planet, contact is made.

"There is a way that we can prevent the global extinction of intelligent life on any given planet. Only a very few require this help. This method in no way affects the outcome of a planet's overall race maturity and freedom to evolve in its own unique way. It is perfectly interwoven in their movement in time and space. You will discover for yourself at a later date the way this is accomplished. It cannot be explained or understood in the language of words or by science as you know it in your time. We just wanted to let you know that we do not always remain passive and let a planet's civilization completely annihilate itself.

"The vast majority of planets make it entirely on their own, as did our Earth, though perhaps not entirely on its own. Many miraculous events took place that cannot be explained in the natural realm. Had it not been for these inexplicable occurrences, we would not have made it.

"If this sounds as if there are no warlike civilizations capable of interstellar space travel, we are glad to report that this is the reality we now live in. It may seem too good to be true, but absolute intergalactic peace exists throughout the Milky Way. Of course, many are going through the same trials and tribulations of growing pains that we went through.

"If ever a civilization reached the point of interstellar travel, and became invasive into the sacred domain of another planet-world, it would be instantly stopped. There would be no contest. Not by any kind of a destructive means but by a complete neutralization of any and all weapons of war that can be used outside the realm of their own planet. There would be no loss of life, nor any disciplinary action. As of this date, an aggression of this nature has never occurred.

"I would like to close with a moment of silent gratitude and homage to you our fellow citizens of 21st Century Earth. We owe so very much to you. You have our promise and commitment that we will carry on the amazing legacy you have passed on to us. That legacy is to honor and preserve all life, wherever found, in whatever form it may appear."

The Meeting

Professor Patton welcomes guests from The Federation of Blue Planets.

AFTER MY LECTURE, I was approached by a fellow colleague.

"Dr. Patton, I have a message for you from one of our Blue Planet members. They ask that you meet them within our communication area, regarding something that you will be very interested in."

"Thank you, Dr. Thomas; I will report at once."

The meeting is held within the University's own enormous atrium facility. This area features a large seven-story window overlooking the city skyline of Albuquerque.

I am graciously greeted by our comrades from this other world. They resemble our basic human form, except they exhibit a more beautiful form and appearance than we. All three, which are two men and a lady, are young, being in the prime of their maturity. Their skin emanates a soft glowing effect, and they radiate a very pleasant and friendly atmosphere.

I welcome them, thanking them for contacting me, expressing my gratitude and excitement for what I consider to be a historical event. This is the first time we, Earth, have made person to person contact with this world. Since they are able to speak our language fluently, they immediately begin to communicate why they are here.

"Thank you, Professor Patton, for your warm welcome. In your language, my name could be translated into the name Logan; likewise, in the same way for my companions Mira and Tyron. The three of us, as representatives of our entire planet, welcome you into our family of Blue Planets. We are aware of the history of your planet, especially the era that your message through time was addressed."

Paradoxical Time Travel

Logan explains to Derrick the meaning of time travel.

"IT MAY INTEREST you to know that we have discovered a way to send messages through time to specific individuals. In its wisdom, the universe allows only contact with individuals, not with the masses. We can send your message to your family ancestor, Robert Berg. Then we can follow up with an actual in-person contact. Would you be interested and feel comfortable with this? It is meant to be a welcoming gift from us to you. As you know, in several of his lectures, Professor Berg specifically issued invitations to a future generation of Earth to pay him a visit when time travel becomes possible. He clearly stated in several of his lectures that he would welcome this event with open arms."

I replied, "Interested is not the word. Fascinated and ecstatic is a better way to express my reaction. I have heard this is possible in theory; however, I have questions about the inherent paradoxes in which we all are familiar. Please feel free to expound on this subject. Take as much time as you like."

Logan nodded, "Very well. Let me explain what we mean by time travel. We have discovered that time has a built-in quantum fluctuation that will prevent any paradoxical event from taking place. You are aware of the post-selected model of time travel which involves distorted probability close to any paradoxical situation. Time and space continuum has a built-in safeguard that prevents any occurrence that will change the future in any meaningful way. This safeguard is precisely what Professor Berg included in his theory on the possibilities of time travel. He was right.

"Within the boundaries of this quantum fluctuation, we are still able to move freely, experiencing different eras in time. However, we can only physically step into another time dimension by remaining connected to our own time and space flow. This acts as a protective environment. Our bodies are compatible only within the parameters of the time and space movement in which we are born. Therefore, our physical bodies traveling in time remain connected to our own time continuum. This is accomplished through the application of a science that operates in the fourth-dimensional realm.

"All of this is made possible by the application of an inherent universal law that transcends technology. In the not too distant future, you will be more prepared to understand this science. You will be visible and tangible to the person you have made a time connection with. This will include a random selection of others. We have no control over the random selection which can consist of any living thing. The selection is controlled by and embraced in this universal safeguard that I spoke of earlier. We are very grateful to this apparent universal intelligence and law that maintains the time and space integrity of our cosmos.

"At whatever point in the time-space flow we find ourselves, we can only go back from that point. We cannot travel into the future. However, the future is really at hand here and now with your contact with thousands of other civilizations that are far more advanced than you. They can share with you whatever your planet's culture is capable of understanding and utilizing in a constructive way.

"Any civilization that reaches the place in conscious evolution and knowledge that our galaxy alliance of planets has achieved is experiencing this universe where the present virtually includes the future. It now becomes primarily about where we are in the evolution of Consciousness than where we are in time and space. In other words, when a certain pinnacle of Consciousness is reached, it transcends time and space.

"Time travel is engaged in on rare occasions and only for significant personal reasons. For the most part, there is little interest. Those who are capable of time travel are already living in a paradise existence far beyond anything that could be found in the past. Our galaxy alliance of planets has evolved to the place of infinite possibilities right here in the present. When this point is reached, the future and the present become one."

Beyond the Science of Technology

Logan expounds on Professor Patton's intense interest and passion about the civilizations that have transcended technology.

"SOME CIVILIZATIONS have transcended an existence that is limited to technology. We, ourselves, are in the beginning stages of this transition. These are the ones that have so captivated your interest. If you would like, I can briefly expound on this."

My response was an enthusiastic, "Please do so! You already know how deep my passion is about this subject. Feel free to treat this as a teacher and student relationship."

Logan replied, "Very well. Beyond our time and space cosmos, there are multitudes of other time and space dimensions, most of which are much like ours. As you well know, there is a great deal of scientific evidence and proof of this. They all appear to be interconnected, unfolding as a multidimensional entity. The cosmos that we inhabit is just one of a vast multiplicity of universes that make up the all-encompassing family of universes.

"This family of universes is amazingly compatible in supporting the same type of life forms. In other words, as in our cosmos, we could live and breathe on many of their planets the same as we do here. The laws governing our neighboring universes are essentially equivalent to ours. In fact, many of these universes have far less intergalactic chaotic conditions.

"All of these time and space dimensions are interconnected by a vast network of portals that serve as doorways from one to the other. We did not create these portals; we only discovered them. From within our cosmos, we have access to all of these other universes. A few civilizations in our galaxy have already visited several of them. The reports are that many are far more beautiful, colorful, and exotic than ours. Intergalactic harmony seems to be the predominant law governing these other dimensions.

"If it is okay with you, Professor Patton, I will be happy to continue this subject at another time. I know that you are greatly anticipating your contact with Robert Berg. After your contact with him, I will provide you with some profound insight into a teaching and science that we both find immeasurably fascinating. It is about an astonishing and remarkable adventure in a new kind of Science that is coming from a few of the Blue Planets that have transcended the need for technology. In your language, this Science has been given the name, Transcendental Science."

A NEW DISPENSATION IN SCIENCE

The astonishing and remarkable adventure in a new kind of Science

"THIS NEW DIMENSION and dispensation of Science have led to the discovery of a universe that transcends all time and space dimensions, revealing an infinite universe that has no bounds or limitations. In this realm, there is no such thing as space as we know and experience it. The cosmos perceived within the limitations of our material or physical senses provides us with a finite view of the universe. Our Consciousness, however, has unlimited potential to encompass and apprehend the universe as it actually is.

"This universe, which transcends time and space dimensions, is teeming with an infinitude of worlds of indescribable beauty. The wonders and majesty of Nature manifest a loveliness, splendor and profound tangibility far beyond anything witnessed in our time and space dimensions. This realm can only be entered into through an evolution of Consciousness that transforms an individual into a higher expression of life. This higher expression of life also includes a transformation of the body that makes it compatible to exist in this realm. Individuality will forever be maintained."

A Message Through Time

Logan explains to Derrick how they will send a message through time as a prior step before actual person to person contact.

"WHAT YOU ARE ABOUT to see, Professor Patton, will appear only a few feet away from you. It will be an opening in the time-space continuum. This is the portal in time in which you will deliver your message and, after that, step through to actually stand face to face with Doctor Berg. When the portal opens, you will be viewing Robert Berg in the year 2042 December 24, Christmas Eve at his home in Albuquerque, New Mexico. At this date, Robert is 61 years of age."

The time flow starts at 9 pm. The time portal opens. I am now witnessing Professor Berg, along with his wife, their children, and grandchildren, standing in front of a large and beautifully decorated Christmas tree. They are all gathered around a Grand Piano singing Christmas carols. Their large two-story great room window gives a beautiful viewing of a light snowfall in the background. Apparently, the entire family is there for the holiday season.

The size of the time portal opening is the very same size of the entire Room. I am stunned and captivated by this incredible event. I watch in joyful celebration with them for about an hour. His grandchildren are now being put to bed eagerly anticipating Christmas morning.

Professor Berg has just entered his study and library and is now sitting in front of his computer.

Logan said, "Derrick, probably the best way to present your Federation message is for it to appear as if it were on or somehow connected to Dr. Berg's computer screen. However, this will not be connected to his computer in any way. It will be obvious that it is coming from a source outside of his internet communication system in that era. This will be in a three-dimensional format that will display very unusual colors that he has never seen before.

"After we see his reaction to the message, then you can decide as to whether to go forward with the next step. The next step will be you speaking into the portal which will be translated into the same computer format as the previous Federation message. This is to give him notice that you are about to pay him a personal visit through time, contingent upon him granting you permission to do so. He will probably be suspicious that this is all some kind of a hoax, not really believing this would actually occur.

"I would suggest that we give him the option of how you are to appear. Out of thin air so to speak or another way such as a knock at the front door for an initial greeting.

Then after the initial meeting, we can provide him with conclusive proof. Through the time portal, he can see our world but, as I have mentioned before, he cannot step through into the future."

As the message was appearing to Professor Berg, his reaction was that of being startled and very puzzled. He became completely still and extremely focused on the message. After a few minutes, he seemed to be even more baffled. That is when I started to speak into the portal to try and provide some additional proof that this was not a cyber attack on his computer. Being aware of his well-known and acclaimed theory on why time travel is possible within the boundaries of a quantum fluctuation law, I knew that the mention of the subject would immediately catch his attention.

My verbal message was being converted into the same type of word design and configuration as the other. This included parts of Professor Berg's personal diary that had been handed down to me. None of this information was known at that time by anyone but him. He was now reading my request asking for permission for a personal visit by either stepping directly through the time portal into his study, or some other way, or not at all. Just to let me know. Whatever your decision, it will be respectfully honored.

THE REUNION

**Spanning almost 500 years,
Derrick and Robert meet face to face.**

HE HESITATED for a moment and said with a sharp, authoritative voice as someone who is calling your bluff, "Okay, step straight through!"

As I stepped through, the entire time portal opening became visible to him. He could now see into the 25th Century. When I came into view, along with the background scenery of my world, Robert reacted with stunning surprise and excitement. There seemed to be no sign of fear or confusion on his face. The vision he saw through the time portal was the Albuquerque City skyline from the University Atrium in the twilight of the evening. The look on his face as he witnessed this City of the future was one of awe and astonishment. He saw into a world of such magnificent architectural design and color that it seemed to him a city of pure light made tangible. This City was set in an incredible naturist vista of beauty that defies description.

His first words were, "Oh, my God, can this be real?"

My prior message had somewhat prepared him to know who I was. I could still see that he was in a state of disbelief which was perfectly understandable.

I greeted him with, "Professor Berg, this is indeed an honor to meet you. Thank you for allowing me into your home. I realize that a profound event such as this is not easily taken in so quickly. I feel the same way also. This is the first time that anyone from Earth has ever traveled through time. This is a historic occasion for both of us. By the way, Merry Christmas to you and your family!"

Professor Berg responded, "It may take me a few moments to accept the fact that this is actually happening. Please bear with me.

As you know, I have publicly stated in many of my international lectures that I am formally giving an invitation to a future generation of Earth to contact me through time if ever this science is achieved.

"Tell me again, Professor Patton, that we made it in the spectacular way in which you described. This has to be by far the best Christmas gift I have ever received. The joy I am feeling at this moment is indescribable. I hope you can stay for awhile. I have so many questions, I don't know where to begin. Since you already know most everything about me, please tell me about you and your family in this wonderful era that our Earth is now experiencing."

I spent almost an hour expounding on how life was now for everyone while Dr. Berg was witnessing 25th Century Earth through the time portal. I watched as he gleamed with joy and gratitude with every word I was telling him about a future reality far greater than he could ever have hoped for, dreamed, or imagined.

Then came a soft knock at the door. The time portal behind me immediately became invisible. Only Robert Berg and I were together alone. Robert seemed to know who was knocking.

He said, "I'll bet that is Amanda and David, my two grandchildren, excited about Christmas and cannot sleep. They did this same thing last year at about the very same time."

Then the sweet voice of a child said, "Grandpapa; may we come in?"

Robert replied with an excited and cheerful voice, "Of course; I was hoping you would come and see your granddaddy before you went to sleep."

They bashfully came in radiating an innocence and cuteness that only the purity of a childlike heart can personify.

Robert said, "I have a special friend and relative I would like for you to meet. He came to bring me a Christmas gift that I will treasure forever. This is Derrick. Derrick, this is Amanda who is 6, and this is David who is 4."

I responded, "I am so happy to meet you, David and Amanda. I see that you both are having a wonderful Christmas."

Amanda asked if I were going to stay and watch them open Christmas gifts in the morning. I thanked her for asking me to stay but said that I had to leave tonight to return home and be with my family. I was sure this would be the beginning of many trips that I will be making from this time forward. I hoped we would see each other again soon.

Robert lovingly suggested to his two grandchildren that they go to bed and say a prayer for boys and girls all over the world. Wishing them a Christmas filled with love and happiness.

Amanda and David gave Robert a hug and a kiss on the cheek. Both said, "I love you, Grandpapa. Merry Christmas."

After David and Amanda had left the room, I informed Robert that I could make more of these private visitations with his permission.

He quickly responded with an enthusiastic, "Of course, absolutely. I was hoping this would be the beginning of an ongoing friendship and family connection. A connection with the knowledge, peace, and absolute assurance that my family and all of humanity has a magnificent future ahead. And what an amazing family connection it is!"

My farewell words to Robert were regarding giving him advance notice of my next visit in the same way as this one. He would see this same type of text message that would be visible only to him. We embraced each other with tears of joy.

Robert looked straight into my eyes and said, "Derrick, I love you. I could say that you have no idea what all of this means to me, but I am sure you do. I have been lifted to a place of inspiration and vision that words are inadequate to describe. Because of this experience, I am sure that my capacity to be even more dedicated in playing my part in establishing world peace will be greatly enhanced. My knowing the outcome will only motivate me to pursue this sacred quest for Earth's destiny with a spirit of joy and celebration."

I responded, "You can be sure that my life will never be the same. Just as you have described, Robert, I have been lifted into an insight and inspiration that I have never before experienced. Your comment about you knowing the outcome apparently has had no effect on the future outcome. Therefore, when I return, I will present this question to my friend Logan. I think I know why, and I am sure Logan already knew the answer to this all along."

Robert smiled and replied, "I am interested in this idea of yours, Derrick, that my knowledge and actions in this time period did not or could not change the future."

I responded, "I have a theory on this; however, I cannot be sure until I confer with Logan on this matter. I believe that this event of my coming here may already be accounted for in time. Any other trips will be the same accounting for in time regardless of when, or whatever event or circumstance leads up to my visit.

"You were correct, Robert, regarding your quantum fluctuation theory you so elegantly put forth in lectures and essays. This time and space continuum has a built-in safeguard that absolutely prevents any occurrence that will change the future in any meaningful way.

"I wish I could stay longer, but Logan tells me it's time for me to return. I am looking forward to seeing you again soon. Farewell, my dear friend."

Robert replied, "You are always welcome here, Derrick. Please visit as often as you can. I hope to see you soon."

When I returned, I confided with Logan about my incredible experience.

****

CONTINUING BOOKLETS of "The Federation of Blue Planets" will follow Derrick Patton's ongoing visits with Robert Berg. This warm and loving family bonding develops and grows into a relationship that transcends the bounds of time and space.

Readers will be taken on incredible adventures of beauty and wonder in Earth's exploration of the Blue Planets.

It will also include Logan revealing to Derrick a greater understanding of this Transcendental Science that is opening up a gateway into a realm beyond technology.

Josh David Hamilton

www.ingramcontent.com/pod-product-compliance
Lightning Source LLC
Chambersburg PA
CBHW061204180526
45170CB00002B/960